Bee HIVES AND Bat CAVES

Amazing Animal Homes

From the Editors of OWL Magazine

Edited by Katherine Farris

Illustrated by Andrew Plewes

Greey de Pencier Books

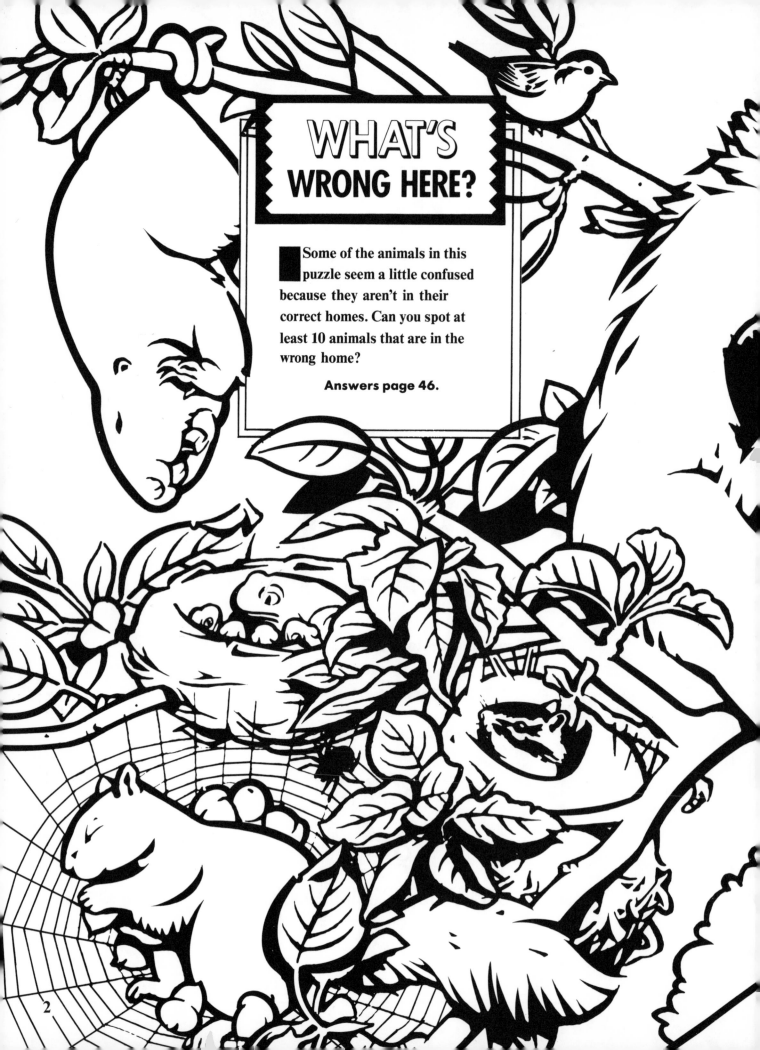

WHAT'S WRONG HERE?

■ Some of the animals in this puzzle seem a little confused because they aren't in their correct homes. Can you spot at least 10 animals that are in the wrong home?

Answers page 46.

3

JUNGLE PUZZLERS

Jungles around the world are home to many strange animals, some so weird it's hard to believe they really exist. Look closely at the unusual creatures on these two pages. Most of them are real, but two aren't. Can you tell which are which?
Answers page 46.

2. Asian Mount Omei Treefrog

Suction cups on my hands and feet help me cling to trees. I lay eggs in a protective foam on tree twigs overhanging a pond. When my eggs hatch, the tadpoles fall into the water.

Am I bluffing you?
☐ YES ☐ NO

1. The Soldira Butterfly from Malaysia

During heavy rainstorms, I stick my feet to a tree branch with a slimy glue. After the storm, I unstick my feet with another substance and off I fly.

Am I bluffing you?
☐ YES ☐ NO

3. The Petal-Faced Bat from South America

That frilled "flower" surrounding my nose acts like a dish antenna to pick up faint signals from flying insects. When I sleep, I close my "petals" and my large ears over my face.

Am I bluffing you?
☐ **YES** ☐ **NO**

4. The Aye-Aye from Madagascar

With my keen sense of hearing I find larvae and insects under dead bark. Then I use my curved front teeth to gnaw a hole in the wood. With my narrow middle finger I hook out my prey and pop it into my mouth.

Am I bluffing you?
☐ **YES** ☐ **NO**

5. The Flower Mantis from Tropical Asia

I'm disguised to look like the orchids I live among. But insects beware! Underneath those "petal" covered front legs are some not-so-pretty spines. Once I've caught my dinner, I won't let go.

Am I bluffing you?
☐ **YES** ☐ **NO**

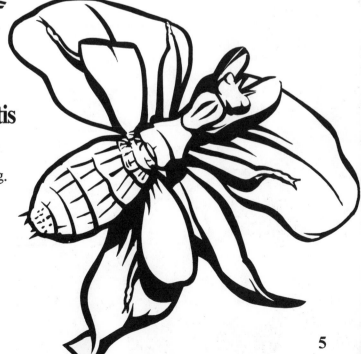

HOT, HOT, HOT CROSSWORD

Here's one way to bear the heat. Try to figure out the answers to all the clues in this desert crossword.

Answers page 46.

Across

2. Desert days are hot but nights can be _ _ _ _.

5. Biggest desert in the world

6. Spadefoot toads lay their _ _ _ _ in pools of water.

7. Six-legged animals with thick skins for desert survival.

8. Opposite of wet

9. Place in the desert with water and trees

10. What a snake does to a mouse running by.

12. Desert _ _ _ _ _ _ store water in their leaves, trunks and spines.

14. Very little _ _ _ _ falls in the desert.

16. Desert plants have shallow _ _ _ _ _ so they can absorb rain as soon as it falls.

20. Insects carry it from one plant to another.

21. _ _ _ constrictor.

22. _ _ the light of the moon.

23. People who live in the desert and move around it a lot

24. You and _ _.

25. I spy with my little _ _ _.

Down

1. The only animals with humps on their backs

2. A desert plant

3. This crossword is about the _ _ _ _ _ _.

4. The world's biggest bird

5. A frog breathes through its nose and _ _ _ _.

9. These birds fly at night.

10. The Sahara desert is on this continent.

11. Just like you, desert animals need water or they get _ _ _ _ _ _ _.

13. Cactus leaves are called _ _ _ _ _ _.

15. Opposite of yes

17. This desert animal has no legs.

18. Not closed

19. This insect makes cats or dogs itch. One type lives in sand.

21. Insect that makes honey

22. Rhymes with 24 across.

CREATE A CREATURE

What kind of animal do you think would live here? There's a tall tree to sleep in, fruit and nuts to munch, insects to catch, a grassy patch to lie in and a river to cool off in.

Draw an imaginary creature that you think would make its home here.

NEAT NESTS

Birds build their nests in many different places and use all sorts of materials. Can you match each of these birds with its home?

1. Barn Swallow

I plaster my cone-shaped nest to walls, especially under eaves.

2. Ruby-throated Hummingbird

My nest is so tiny that I bind it together with spiders' webs.

3. Barred Owl

Check for pine sprigs in a hollow tree stump—that's my nest.

4. Red-winged Blackbird

My nest hangs between two strong reeds in a marsh.

5. Red-tailed Hawk

Look for my large, flat nest high up in a pine tree.

Answers page 46.

BUSY BUZZ

■ Use these close-up illustrations to help you fill in the blanks and find out about this animal and its home.

The _____ buzzes around gathering nectar from _____. It carries the food in sacs on its _____ back to the _____ where the food is made into _____.

Answers page 47.

WHOSE HOME IS IT?

red = 5 − 4 + 0 =

pink = 3 + 4 − 5 =

orange = 2 × 2 − 1 =

brown = 9 − 7 + 2 =

black = 8 − 4 + 3 =

■ Can you identify this animal home? First, solve each of these equations to find out what number each color is. Then, use the numbers to help you color in the picture and guess whose home this is.

Answers page 47.

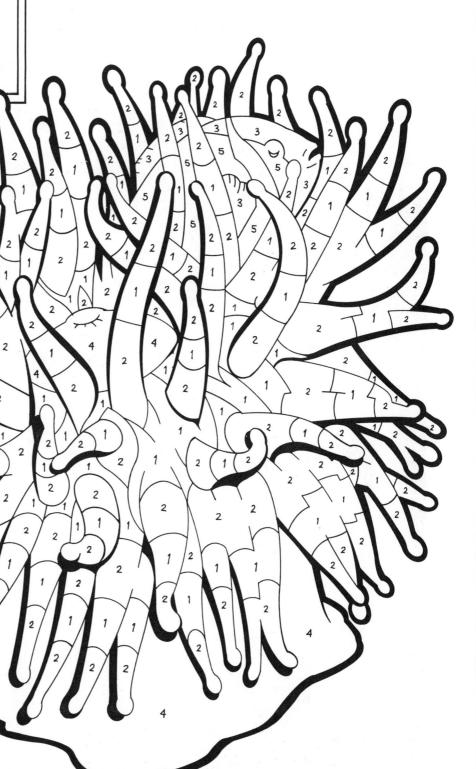

MIGHTY MITES
MEET LEAPING DUCKS

Nick, Sophie and Mark Mite are three special kids with a big secret: they have discovered a way to shrink to any size they want and grow big again. The Mites are bird-watching when a brightly colored duck catches their attention.

OH WOW! LOOK AT THE COLOURS ON THAT DUCK! GREEN, BLUE, BRONZE...

WHERE? LET ME SEE!

DO YOU MEAN THAT ONE SITTING ON THE BRANCH?

WHAT BRANCH? LOOK IN THE WATER.

WELL, I CAN SEE DUCKS SITTING IN A TREE.

DUCKS IN TREES?

YOU'RE RIGHT, NICK. THERE'S A DUCK STICKING OUT OF THIS TREE.

THE MITES SHRINK

LET'S SHRINK AND GET CLOSER.

THE MITES HAVE FOUND SOME WOOD DUCKS. FEMALE WOOD DUCKS NEST IN TREE CAVITIES; THE MALES GATHER ON NEARBY BRANCHES.

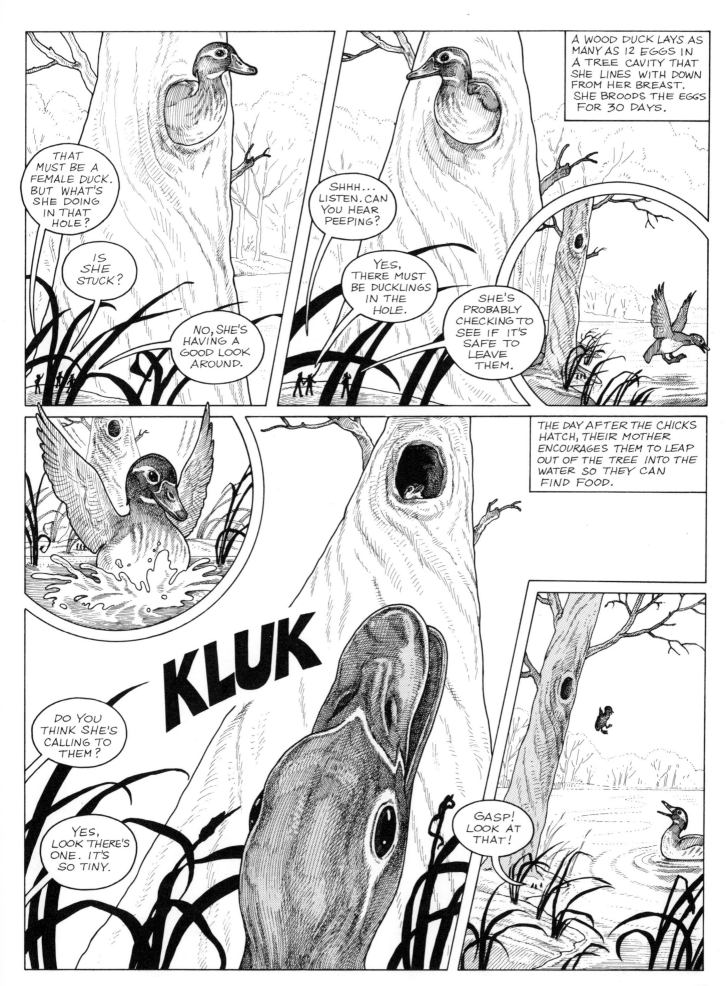

A WOOD DUCK LAYS AS MANY AS 12 EGGS IN A TREE CAVITY THAT SHE LINES WITH DOWN FROM HER BREAST. SHE BROODS THE EGGS FOR 30 DAYS.

THAT MUST BE A FEMALE DUCK. BUT WHAT'S SHE DOING IN THAT HOLE?

IS SHE STUCK?

NO, SHE'S HAVING A GOOD LOOK AROUND.

SHHH... LISTEN. CAN YOU HEAR PEEPING?

YES, THERE MUST BE DUCKLINGS IN THE HOLE.

SHE'S PROBABLY CHECKING TO SEE IF IT'S SAFE TO LEAVE THEM.

THE DAY AFTER THE CHICKS HATCH, THEIR MOTHER ENCOURAGES THEM TO LEAP OUT OF THE TREE INTO THE WATER SO THEY CAN FIND FOOD.

KLUK

DO YOU THINK SHE'S CALLING TO THEM?

YES, LOOK THERE'S ONE. IT'S SO TINY.

GASP! LOOK AT THAT!

NIGHT VISITORS

During the night this garden had several visitors. Look at the animals and their tracks at the bottom of the page to see who came to the garden. Then play the true or false game to find out what they did.

1. A badger came out of the forest and walked to the apple tree.

☐ **TRUE**   ☐ **FALSE**

2. A rabbit ran into the vegetable garden.

☐ **TRUE**   ☐ **FALSE**

3. A fox caught the badger.

☐ **TRUE**   ☐ **FALSE**

4. A deer visited the bird feeder.

☐ **TRUE**   ☐ **FALSE**

5. A rabbit ran from the vegetable garden to its burrow.

☐ **TRUE**   ☐ **FALSE**

6. An owl caught a squirrel.

☐ **TRUE**   ☐ **FALSE**

7. An otter slid down the mud bank into the lake.

☐ **TRUE**   ☐ **FALSE**

8. A mouse ran from the bird feeder to the house.

☐ **TRUE**   ☐ **FALSE**

Answers page 47.

SPLASH!

What do you think this swimmer will see as she dives down to the bottom of the lake? Draw in the animals and plants that you think live there.

WHO'S HIDING HERE?

In thick jungles, many animals rely on camouflage for protection. Some have striped or spotted coats so they can hide among the jungle's tall grasses and trees. Some animals rely on dull colors to hide them. Others are brightly colored so they blend into the jungle's flowers and plants. Can you find the animals hiding in this jungle?

Answers page 47.

PRAIRIE
DOG RUN

Can you help this mother prairie dog find her way through this maze of underground tunnels and home to her babies?

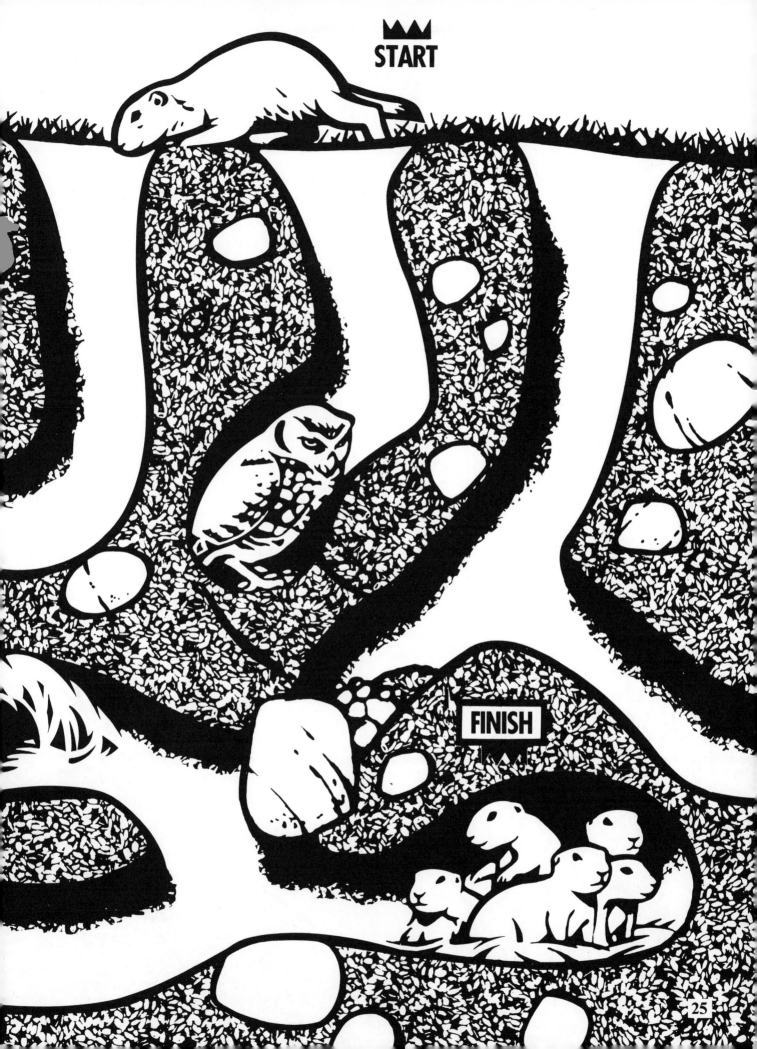

START

FINISH

25

EATING IN

Some animals don't even have to leave home to catch their dinner. Who are they? Find out by matching each clue to an animal in the picture.

1. My silky web looks delicate but it's strong. When my prey gets caught on the sticky threads, I roll it up in a bundle to eat later.

2. With my strong hind legs, I dig a funnel-shaped pit in sandy earth. Then I burrow down so that only my head and open jaws can be seen. I sit motionless until an ant or other small insect falls into my mouth.

3. I hang from tree branches in tropical rainforests of Central and South America. I don't have to move to eat since I munch on the leaves, blossoms and fruit that are close by.

4. I stick my long, feathery legs out of my shell and wave them through the water to trap bits of food.

5. I do eat out in summer, but in the winter, when ice covers my pond, I stay snug at home and munch on branches I've gathered.

Answers page 47.

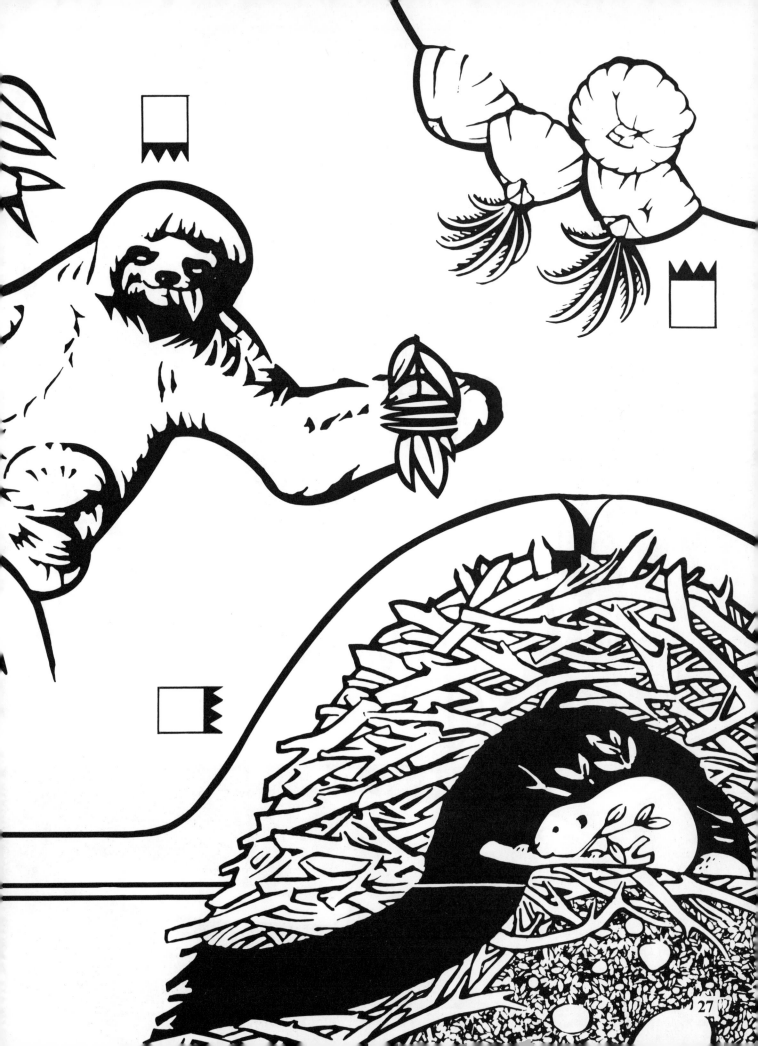

COZY CAVES

Caves make cozy homes for animals known as troglodites. "Trogs" have lived for so long without light that they are blind and ghostly pale. We've put trogs from all over the world into this one cave—and, just to fool you, we've included one bluffer. Can you figure out which animal would never live in a cave?

1. This tiny bat lives in North American caves. It sleeps in the cave at night and hibernates there through the winter. Its droppings are an important source of food for other cave dwellers.

2. Waitoma glowworms cover cave roofs in New Zealand and Australia and twinkle on and off like Christmas lights. These glowworms are actually fly larvae living inside slimy, open-ended sacs. They spin sticky 2 m (6½ ft) threads that dangle in the dark and trap small insects.

3. Because of the cold and shortage of food, the feathery gilled, blind salamander of Yugoslavia grows three times more slowly than other salamanders. When food is really scarce, it can survive for a whole year without eating.

4. Water insects beware! This hungry fish from the caves of Kentucky doesn't need eyes to hunt. It can feel even the tiniest movement in the water through special cells in its skin.

5. The small, see-through blind crayfish of Florida is so light that it can hang from the surface of the water. While there, it cleans its long legs and antennae, removing bacteria that it probably eats.

6. This black and red bee lives deep in caves on the windy western side of the Rocky Mountains. In the winter, it eats algae that grow on the cave walls. In summer, the bee flies out and fills up on nectar to make its special cave honey.

Answers page 47.

WHO'S UNDER THE SNOW?

Have you ever looked at a field covered with snow and wondered what happened to all the animals that lived there in the summer? In this winter scene, all you can see are two sparrows feeding. But don't be fooled. There's a lot going on— under the snow! Draw what you think is happening under this snowy blanket.

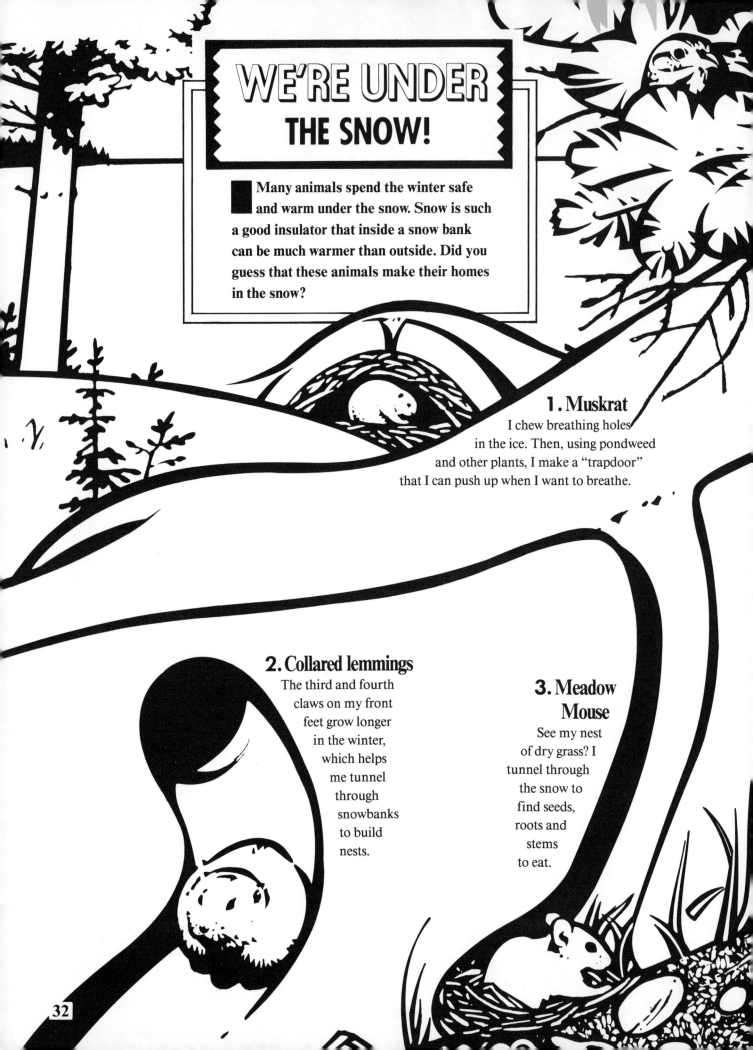

WE'RE UNDER THE SNOW!

Many animals spend the winter safe and warm under the snow. Snow is such a good insulator that inside a snow bank can be much warmer than outside. Did you guess that these animals make their homes in the snow?

1. Muskrat
I chew breathing holes in the ice. Then, using pondweed and other plants, I make a "trapdoor" that I can push up when I want to breathe.

2. Collared lemmings
The third and fourth claws on my front feet grow longer in the winter, which helps me tunnel through snowbanks to build nests.

3. Meadow Mouse
See my nest of dry grass? I tunnel through the snow to find seeds, roots and stems to eat.

6. Snowfleas

We may look like pepper sprinkled on the snow but we're called springtails, because of our habit of suddenly "springing" forward in the snow. Look for us in late winter on the south side of trees.

5. Ruffed Grouse

I keep warm by folding my wings and diving headfirst into a snow bank. There I make small tunnels to find plants to eat.

4. Vole

I make highways through and under the snow. But I also make runways up and along branches and trunks of woody plants and trees so I can eat the bark.

HILLS AND HIVES

■ Can you find the names of these animals and their homes in this wordsearch? To make it tricky, we've hidden the animals' names separately from the names of their homes. We've done the first one for you.

Answers page 47.

```
W R A L L I P R E T A C
O H A O A T S N A I L O
R M N N H B E A V E R C
R E T I B B A R P I T O
U S L O D G E W M H S O
B L I O T T E R E I E N
B F O X E D R E Y V T S
E A N H O L T E T E T E
E H D O M O U N D E N L
B E W G M T S E N I E O
L E S A E W D R I B P H
S Q U I R R E L L E H S
```

ANT HILL	OTTER HOLT
ANTLION PIT	RABBIT BURROW
BADGER SETT	SNAIL SHELL
BEAVER LODGE	SPIDER WEB
BEE HIVE	SQUIRREL DREY
BIRD NEST	TERMITE MOUND
CATERPILLAR COCOON	WEASEL HOLES
FOX DEN	

Now, moving from left to right and then top to bottom, write the letters you didn't use in the blanks below to discover a phrase you and the animals might use to describe your homes:

_ _ _ _ _   _ _ _ _ _ _ _
_ _ _ _ _

34

SCRAMBLED NEST

Uh-oh! Someone dropped these photos of a bird making a nest and now they're all mixed up. Can you put them in the right order?

Answers page 47.

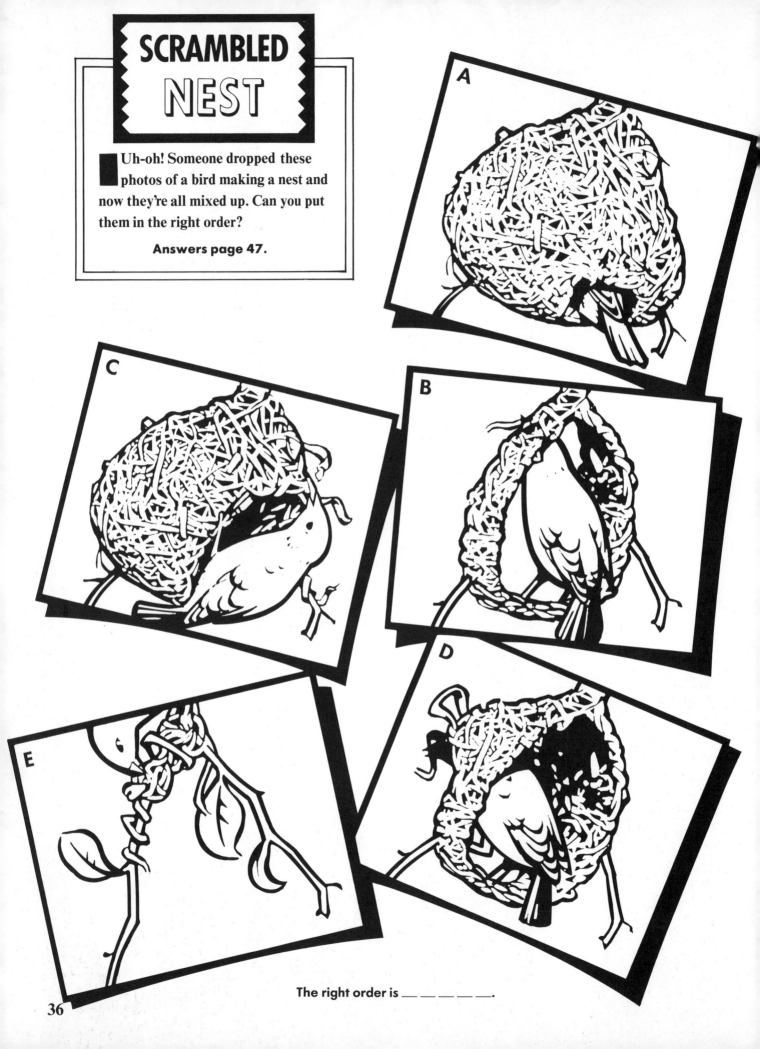

The right order is ___ ___ ___ ___ ___.

MAKE A SNAKE

Snakes live in rock crevices and come out to bask in the sun. But here's a snake that doesn't care what time of day it is.

You'll need:
- an old neck tie
- needle and thread
- scissors
- dried beans or uncooked rice
- a small piece of red paper or felt
- two buttons

1. Sew the thin end of the tie closed.

2. Stuff the tie with beans or rice.

3. Cut a snake tongue from the paper or felt. Sew or glue it into the tie's open end.

4. Sew the open end closed.

5. Sew on two buttons for eyes.

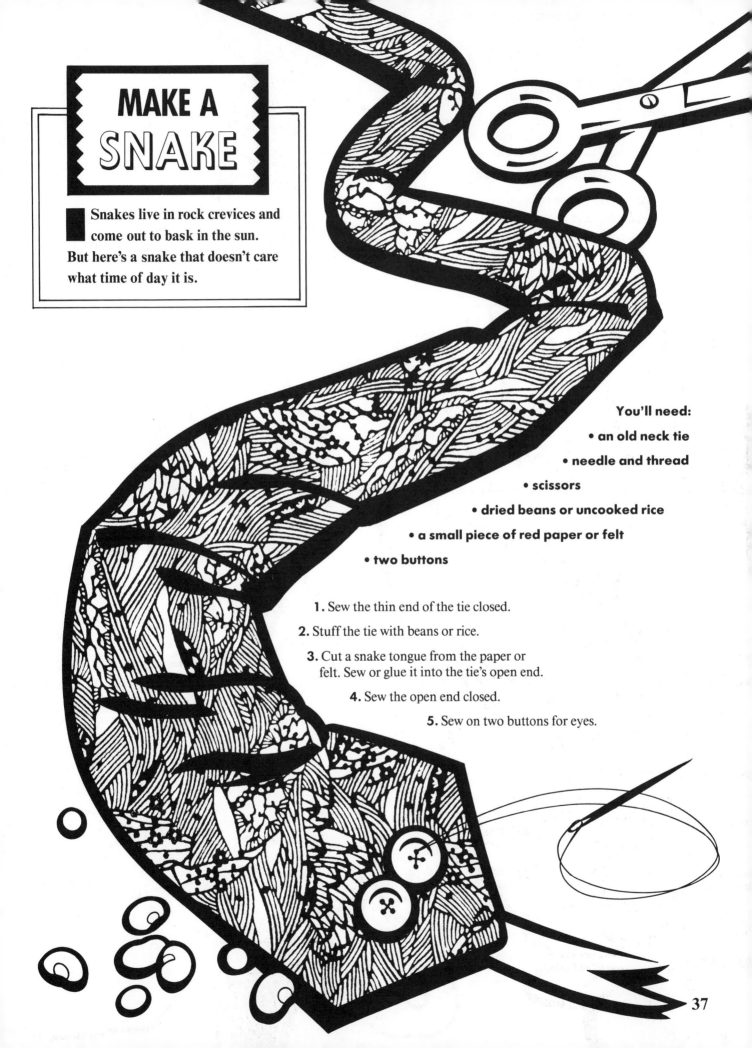

MIRROR LAKE

A calm lake reflects everything around it perfectly, right? Look again. Can you spot the reflections here that are different?

Answers page 47.

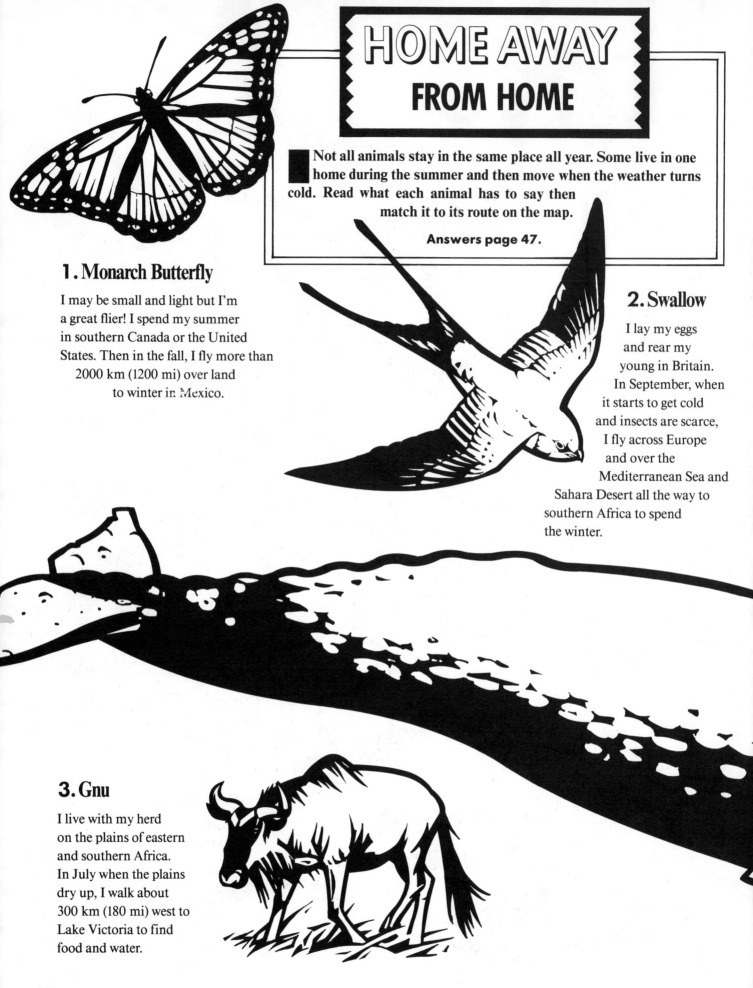

HOME AWAY FROM HOME

Not all animals stay in the same place all year. Some live in one home during the summer and then move when the weather turns cold. Read what each animal has to say then match it to its route on the map.

Answers page 47.

1. Monarch Butterfly

I may be small and light but I'm a great flier! I spend my summer in southern Canada or the United States. Then in the fall, I fly more than 2000 km (1200 mi) over land to winter in Mexico.

2. Swallow

I lay my eggs and rear my young in Britain. In September, when it starts to get cold and insects are scarce, I fly across Europe and over the Mediterranean Sea and Sahara Desert all the way to southern Africa to spend the winter.

3. Gnu

I live with my herd on the plains of eastern and southern Africa. In July when the plains dry up, I walk about 300 km (180 mi) west to Lake Victoria to find food and water.

4. Arctic Tern

I'm the world's champion migrator. During May and June I lay my eggs and raise my young on rocky Arctic ground. When the weather gets cold, I fly 8,000 km (5,000 mi) south. For mid-flight snacks, I dip down to the surface of the ocean to pick up a fish.

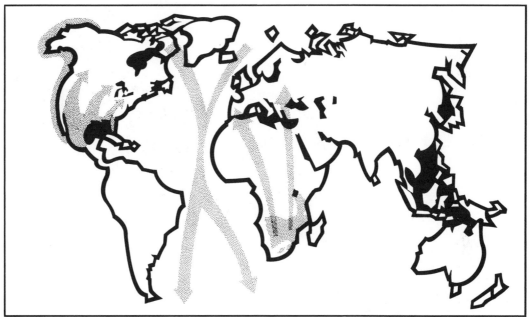

5. Grey Whale

From June to September I live in the north Pacific Ocean eating my fill of tiny shrimp called krill. When the weather gets cold, I head down the west coast of North America to breed in the warm waters off the southern California coast.

HOME SWEET HOME

Here are just a few of the incredible homes where animals live. But beware! Although all the animals are real, one of these homes is imaginary.
Can you guess which one?

Answers page 47.

1. Up in the Air

The praying mantis lives in church steeples in Europe and northeastern North America. During the day, it lies completely still in the belfry waiting to pounce on insects, frogs or young birds.

☐ **TRUE**
☐ **FALSE**

2. Mobile Home

Unlike most crabs, a hermit crab does not have a hard shell of its own. Instead it moves into an empty shell which it carries around on its back. When the crab gets too big for its adopted home, it moves into a more spacious shell.

☐ **TRUE**   ☐ **FALSE**

3. Keep Away

In Europe, the peacock butterfly makes its safe home among bright pink, spiky thistle flowers. If a bird swoops down to snatch up the butterfly, the bird gets poked in the face.

☐ **TRUE**
☐ **FALSE**

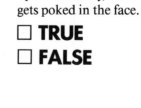

4. Dig In

Tilefish use their mouths and fins to dig burrows
in the ocean floor and in the walls of underwater canyons.
These fish live off the east coast of North America
and their burrows are so tiny that they can't even
turn around in them. They have to swim out backwards!

☐ **TRUE**   ☐ **FALSE**

5. In the Treetop

Orangutans live in northern Sumatra
and Borneo. At night, they climb trees
and weave flat nests of branches to sleep
on. They also snuggle under blankets of
leaves when it's cold.

☐ **TRUE**

☐ **FALSE**

NOW IT'S YOUR TURN

You've just met a lot of amazing animals and seen the incredible places they live. Now try drawing a home for the animal you see here. This creature can burrow in sandy soil, spot danger from far off, climb trees, run fast through fields and hide when danger threatens. What kind of home do you think it would live in?

WHAT'S WRONG HERE, *page 2-3*

Beavers don't hang by their tails from trees. Their tails are wide and flat and they use them as a rudder to steer and to propel them through the water.

Frogs don't lay their eggs or raise their young in birds' nests.

Squirrels don't live in spider webs. Using their tail as a blanket, they curl up in their warm den.

Chipmunks make their homes in underground burrows, not in trees.

Bears don't take naps in trees. They sleep in a cozy, sheltered den.

Owls build their nests in holes in trees. They don't nest on branches like other birds.

Fish don't live in tree holes. But some fish do seek sheltered underwater spots to hide in. Other fish are territorial. They chase away any intruders who swim too close.

Rabbits don't nest in tree holes. Their bedchambers are in underground burrows.

Although they do sometimes live in hollows of trees, raccoons do not share their homes with rabbits.

Eagles don't nest on the ground. They build their nests high up at the top of trees. Sometimes the nests are as big as wading pools.

Seals don't live underground. Their hind feet aren't built for walking on land. But their powerful tail and streamlined body are perfect for swimming. That's why seals stay near the water when they haul themselves up on land. That way they can easily slip back into the water when danger threatens.

JUNGLE PUZZLERS, *page 4-5*
The Soldira Butterfly and the Petal-Faced Bat are the bluffers.

HOT, HOT, HOT CROSSWORD, *page 6-7*
Across: 2. cold, 5. Sahara, 6. eggs, 7. insects, 8. dry, 9. oasis, 10. attacks, 12. plants, 14. rain, 16. roots, 20. pollen, 21. boa, 22. by, 23. nomads, 24. me, 25. eye.
Down: 1. camels, 2. cactus, 3. desert, 4. ostrich, 5. skin, 9. owls, 10. Africa, 11. thirsty, 13. spines, 15. no, 17. snake, 18. open, 19. flea, 21. bee, 22. be.

NEAT NESTS, *page 10-11*

BUSY BUZZ, *page 12*
Bee, flowers, legs, hive, honey

WHOSE HOME IS IT? *page 13*
Did you guess that you colored a giant sea anemone? Some giant anemones grow larger than 1 m (3 ft) across. They are home to the anemone fish, or clown fish. The clown fish is safe in the arms of the anemone. It sleeps and eats there. Sometimes the clown fish leaves the anemone to swim around but if danger threatens, it dashes back home to safety. If the predator follows the clown fish, the anemone curls it tentacles around the intruder and eats it up. The clown fish sneaks bites from the anemone's meal.

NIGHT VISITORS, *page 18-19*
1. False, 2. True, 3. False, 4. False, 5. True, 6. True, 7. True, 8. True

WHO'S HIDING HERE? *page 22-23*
Did you find the toucan, leopard, snake, caterpillar, tree frog, sloth, butterfly, monkey and hummingbird?

EATING IN, *page 26-27*
1. spider, 2. ant-lion larvae, 3. sloth, 4. barnacle, 5. beaver

COZY CAVES, *page 28-29*
The black and red bee is the bluffer.

HILLS AND HIVES, *page 34-35*
The common phrase you should have found is HOME SWEET HOME.

SCRAMBLED NEST, *page 36*
E, B, D, C, A

MIRROR LAKE, *page 38-39*
Did you find these differences in the reflection (starting from the left of the picture)?
The heron is reflected as a flamingo.
The Canada goose is reflected as a seagull.
The beaver is reflected as a rabbit.
The stick is reflected as a water snake.
The deer is reflected as a wolf.
The mountain is reflected as a volcano.
The reflected swallow is flying in the opposite direction.
The owl is reflected as a toucan.
The skunk is reflected as a lizard.
The pine tree is reflected as a palm tree.
Did you find any others?

HOME SWEET HOME, *page 42-43*
The praying mantis is fooling you. It lives throughout North America, northern Africa, Europe and parts of Asia. There it spends most of its life sitting on plants, lying in wait to ambush its prey. When an unsuspecting insect flies by, the praying mantis shoots out its spiny forelegs and grabs its dinner.

HOME AWAY FROM HOME, *page 40-41*

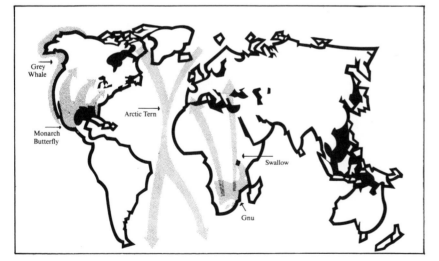

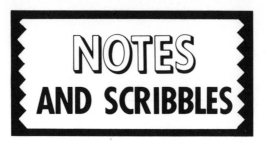

NOTES AND SCRIBBLES